YOUR KNOWLEDGE HAS VALUE

AF131219

Bibliographic information published by the German National Library:

The German National Library lists this publication in the National Bibliography; detailed bibliographic data are available on the Internet at http://dnb.dnb.de .

Imprint:

Copyright © 2016 GRIN Verlag
Print and binding: Books on Demand GmbH, Norderstedt Germany
ISBN: 9783668420328

This book at GRIN:

https://www.grin.com/document/355574

Dr. Marshall Goldberg

Development Of Neocortical Complexity. Synaptogenesis Is Related To The Transformation Of Wolfram Asymmetry Graph #30 To #110

GRIN Verlag

GRIN - Your knowledge has value

Since its foundation in 1998, GRIN has specialized in publishing academic texts by students, college teachers and other academics as e-book and printed book. The website www.grin.com is an ideal platform for presenting term papers, final papers, scientific essays, dissertations and specialist books.

Is the Development of a Fractal Small-World Neural Network Related to the Asymmetry Graphs of Wolfram Cellular Automata #30 and #110?

Marshall Goldberg, MD © 2017

During neocortical synaptogenesis, a collection of developing neurons, representing a random network, organizes itself into a fractal small-world network capable of complex computations.

A problem arises regarding the development of the neocortex because it seems that the number of neurons, synapses, and rules for development of the adult neocortex requires exponentially more information than is available in DNA. Nonetheless, other organ systems such as bone are able to develop complex structures using a small amount of DNA information coded as scale-independent homeostatic negative-feedback rules.

The purpose of this paper is to suggest there may be similar rules for neocortical development requiring a relatively minimum amount of DNA information. These putative rules guide mitosis, cellular differentiation, and synaptogenesis. There is, as yet, no general theory tying together the multitude of complex biochemical events occurring in neocortical development and synaptogenesis. Accordingly, a **heuristic approach** involving the asymmetry graphs of Wolfram cellular automata #30 and #110 is presented. Previous papers [1, 2] describe a method of forming an Asymmetry graph (A-graph) based on the XOR (eXclusive OR) relationships among the digits of the binary number of a Wolfram one-dimensional cellular automaton [3]. It was proposed that,

the A-graph is a scale-free map of information flow in a cellular automaton.

Figure 1 below illustrates the A-graphs of Wolfram rules #30, and #110. Rule #30 (unimodal—one hill) produces intrinsic randomness and uncontrolled information flow. Rule #110 (bimodal—hill-valley-hill) produces complexity and controlled information flow.

The heuristic employed here compares the development of the cerebral neocortex with the conversion of A-graph #30 into A-graph #110 by controlling information flow in the neocortex at intermediate, short, and long distances at all scales.

This implies that the control of excitatory, and in particular, inhibitory synapses is critical in converting a disorganized network into a fractal small-world connectome capable of complex computations [24]*.* It is suggested that:

- the scale-free information flow pattern defined by the A-graph of rule #110 embodies the fractal small-world connectivity and complex information flow in the cerebral cortex;
- the neocortical synaptic connectome is determined by the pattern of early, and ongoing spindle bursts, and neuron transmembrane potentials (Vmem), and;

- the development of excitatory and inhibitory synapses is guided by a negative-feedback homeostatic system.

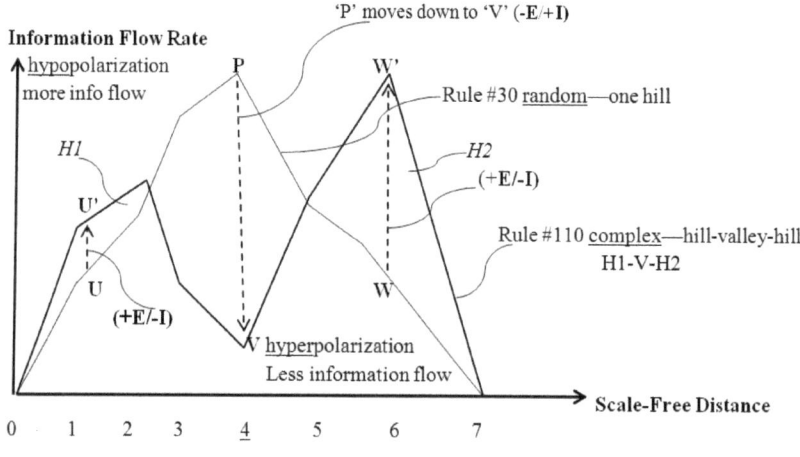

Figure 1

Figure 1 illustrates the A-graphs of rules #30, and #110. The A-graph of a Wolfram one-dimensional cellular automaton is a scale-free map of information flow in that cellular automaton. Rule #30 produces intrinsic randomness, and rule #110 produces complexity. In the figure, (P → V = -E/+I) indicates a decrease in excitatory, and an increase in inhibitory synapses (hyperpolarization = less information flow). In contrast, (U→U,' and V→V' = +E/-I) indicates an increase in excitatory, and decrease in inhibitory synapses, i.e., hypopolarization and an increase in information flow. In converting A-graph #30 to A-graph #110, note that the decrease in area of A-graph #30 at intermediate distances is approximately equal to the sum of the increase in areas of A-graph #30 at small and large distances. If point 'P' in the A-graph of rule #30 moves down to point 'V,' information flow at intermediate distances (at all scales) is decreased, and neuron interconnectivity is decreased[26]. On the other hand, if A-graph #30 shifts upward at small and large distances (U→U' and V→V'), then neuron interconnectivity and information flow is increased at those distances [26]. **The result is that network connectivity is decreased at intermediate distances, and increased at small and larger distances at all scales—the very modifications required to change a random network into a small-world network.** Thus, #30 transforms into #110, with a change from random to complex (controlled) information flow. An analogous process might occur in the developing neocortex where Vmem controls clustering of neurons, such that excitatory synapses (connections) form at small and

2

large distances, and inhibitory synapses (decrease in connections) form at intermediate distances (at all scales). This could result in the connectome changing from a random network to a fractal ultra-small-world network [23]. In a small-world network with N nodes, the distance 'L' between two randomly chosen nodes (the number of steps between nodes) is proportional to the log of N given by, $L = k \ln N$, where k is a constant. Furthermore, in a fractal, scale-free (ultra-small-world) network with significant clustering around 'hubs' we have, $L = p \ln (\ln N)$, where p is a constant. The brain can be considered a scale-free network with significant hubs such as the thalamus, and cortical hubs [17]. This type of connectome allows both a significant decrease in 'wiring' costs, and increased resistance to disruption of the network from random connection losses, unless a major hub is damaged.

Sundelacruz [4] has shown that the cellular transmembrane potential, Vmem, is not only an indicator of cellular activity, but that alteration of Vmem can control cellular functions such as differentiation and mitosis. Changing the value of Vmem externally constitutes a signal controlling downstream biochemical processes associated with these important cellular functions. If Vmem is hyperpolarized, the cell becomes quiescent. At the other end of the scale, if Vmem is hypopolarized, then the cell tends to undergo mitosis. If Vmem has a mid-range value, the cell differentiates and forms synapses. Therefore, it is likely that **external signals control synaptogenesis by controlling Vmem.**

Figure 2, as a general simplification, illustrates a range of Vmem values in cells. Quiescent cells have hyperpolarized transmembrane potentials whereas differentiating cells have a Vmem at some intermediate value, and dividing cells have even lower Vmem values. Cells with even lower Vmem values may undergo cell death (apoptosis) or revert to primitive undifferentiated stem cells, some of them exhibiting the uncontrolled properties of cancer cells.

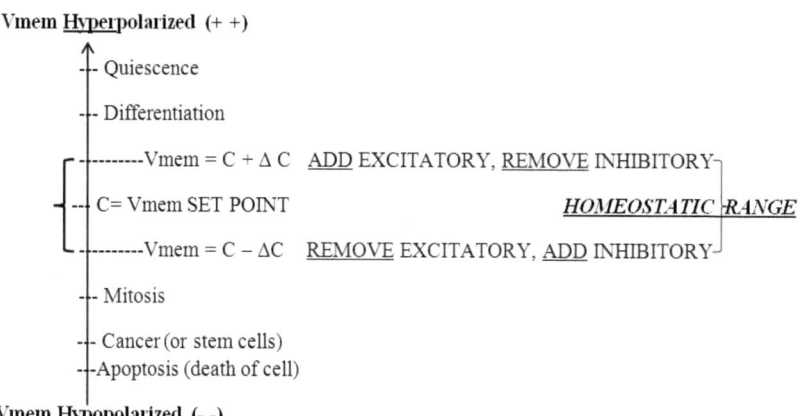

Figure 2

A range of Vmem values and associated cellular behavior is shown. In the fetal brain, developing neurons are subjected to electrical signals from the brain, and from the rest of the fetus. There are random fluctuations (spindle bursts) in the electrical activity of the developing brain, arising from extra-cortical and intra-cortical sources. These random signals are transmitted among cells of the neocortex by both electrical (gap) junctions, and from developing chemical (synaptic) connections[15, 26]. In the 'homeostatic range' noted in figure 2, if Vmem is above or below the set point C, then adjustments to the connectome occur that move Vmem towards the set point. These adjustments rely on negative feedback bio-rules that pare neurons, and adjust the ratio of excitatory and inhibitory synapses. These bio-rules are scale-independent such that the connectome develops as a fractal (ultra-small-world network). Human brain functional networks demonstrate a fractal small-world architecture that maintains global topology while supporting critical dynamics and task-related spatial reconfiguration [17,20]. Note that the adjustments towards C as the set point, do not actually reach C so that asymmetry is maintained in accordance with the rule in a previous paper[1]:

Asymmetry + Computation (Randomization/Entropy) → Complexity.

Thus, conversion of random information flow in Wolfram cellular automaton #30 to complex information flow in Wolfram cellular automaton #110, requires that asymmetry be maintained.

In the early stages of development[16], neocortical information flow is relatively uncontrolled, and exhibits random information flow as illustrated in the A-graph of rule #30 (figure 1). This disorganized information flow correlates with a random network (see Figure 4). However, as inhibitory and excitatory synapses are adjusted according to Figures 1 and 2, then random

4

neocortical information flow gradually becomes complex resulting in an ultra-small-world network analogous to the transformation of A-graph #30 into A-graph #110. Excitatory synaptic connections at intermediate distances 'V' are either eliminated or replaced with inhibitory synapses, and short and long-distance excitatory connections are increased as shown at 'U' and 'W' (see Figure 1). Wolfram class 2 cellular automata (rule #7) produce regular patterns, class 3 (rule #30) random patterns, and class 4 (rule #110) complex patterns. Class 4 is balance between class 2 and 3, between order and randomness.

The adult neocortex is a spatio-temporal fractal, where both the physical geometry of the connectome, and its dynamics are fractals. Because these putative biological 'rules' are considered to operate at all scales, the homeostatic process described in this paper produces a scale-free result.

Self-organization of the neocortex likely arises from the earliest electrical activity of the neocortex. A map of the electrical activity of the developing neocortex over a time interval shows a complex and changing distribution of various transmembrane potentials spread across the neocortex as illustrated in Figure 3 below.

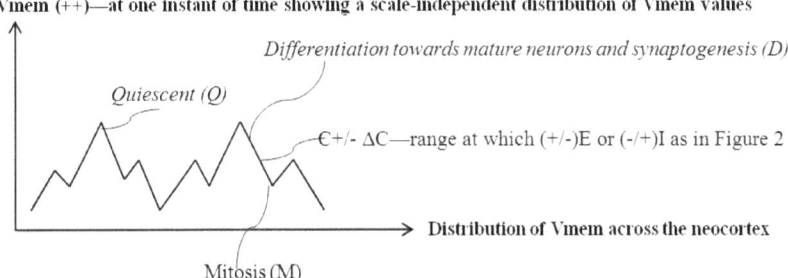

Vmem (++)—at one instant of time showing a scale-independent distribution of Vmem values

Figure 3

Figure 3 illustrates the distribution of Vmem across a two-dimensional slice through the developing neo-cortex at one instant of time. Note that the shape of the curve is postulated to be a fractal, showing similarity at various scales. High points (Q) on the curve represent hyperpolarized, quiescent neurons, and low points on the curve represent hypopolarized neurons (M). Intermediate Vmem values (C+/- ΔC) represent a homeostatic range for neurons that are undergoing synaptogenesis (Figure 2).

According to Sundelacruz [4], the lower points on the curve in Figure 2 represent neurons that are dividing rather than differentiating, i.e. not immediately contributing to synaptogenesis. One may visualize the analogy of variations in the topography of the surface of a pond during a

5

rainstorm. Wave interference can produce <u>standing wave patterns (Seiche)</u>[18]. These standing waves have harmonic wave patterns, i.e., they are fractals. In water, gravity seeks to restore the horizontal surface of the liquid, because this represents a configuration that is in <u>hydrostatic equilibrium</u>. In a similar way, <u>developing neurons seek to restore an equilibrium, i.e., where Vmem values approach C</u> as shown in Figure 2. <u>Spindle bursts</u>[15] in the developing neocortex are both <u>amplitude and frequency-modulated signals</u> arising from various parts of the fetus as well as from the developing brain itself. These spindle bursts can interfere with one another, producing <u>standing Vmem waves in the neocortex</u>. <u>It is hypothesized that the distribution of Vmem values across the neocortex also forms a fractal pattern of standing Vmem waves similar to a Seiche</u>. The transformation of A-graph #30 to A-graph #110, suggests that <u>it is these early signals, and changes in Vmem, that transform the neocortex from a random network into a complex ultra-small-world network by **controlling the distribution and ratios of excitatory and inhibitory synapses at all scales**</u>. As with a Seich, synaptogenesis results in Vmem equilibrium over the whole network. Self-organized critical dynamics can lead to the formation of fractal and small-world networks[20]. Schizophrenia and autism may relate to abnormalities in this system [27].

Points C-ΔC on the curve in Figure 3 represent <u>hypo</u>polarized, potentially mitotically active neurons, because <u>they are receiving more inputs per unit time</u>, thereby resulting in a <u>lower net, time-averaged Vmem</u>. If we are dealing with a <u>negative-feedback system</u>, and if these neurons receive <u>inhibitory</u> inputs, then their net, time-averaged Vmem tends to <u>increase</u>, such that mitosis does not occur. Thus, the stability of the network is maintained. In this regard, it is interesting to consider that in the adult brain, the property of 'resisting' low Vmem values may help explain why neurons do not readily repair in contrast to mitotically active tissues such as the epithelium lining the gut and the mouth. Non-REM sleep, perhaps by allowing a higher time-averaged Vmem among cortical neurons, could stabilize synaptic complexity, and suppress unwanted mitotic activity. The result is a decrease in functional connectivity of the network. As discussed below, we have a homeostatic, negative-feedback mechanism, similar to that seen in the development of structural complexity in bone.

As noted above, because the number of neurons and synapses exceeds exponentially the total information in DNA, synaptogenesis probably depends on relatively few DNA-coded rules. Therefore, we might express a biological rule resulting in synaptogenesis as a **scale-free, negative-feedback homeostatic biologic pseudo-code.**

<u>A biologic pseudo-code for a self-organizing neural network might read</u>:

1. <u>FORM</u>[16] the primitive neocortex using minimum DNA code.
2. Define negative-feedback, homeostatic, and <u>scale-independent rules</u> also using minimum DNA information.

6

3. Accept electrical inputs (SPINDLE BURSTS [15]) from outside, and inside the primitive neocortex.

4. While the time-averaged membrane potential (Vmem) of a developing neuron is greater than some value (C + ΔC), add excitatory synapses or delete inhibitory synapses so that Vmem decreases and approaches C.

5. ELSE, if the time-averaged Vmem is some value (C – ΔC), then remove excitatory synapses and/or add inhibitory synapses, and/or pare the cell from the network, thereby allowing Vmem to approach (but not reach) the value C, thereby maintaining necessary asymmetry (imbalance) in the system.

6. LOOP—continue steps '4' and '5' with adjustment of the relative numbers, and the scale-free distribution and ratios of excitatory and inhibitory synapses until a dynamic equilibrium is reached such that the time-averaged Vmem of all neurons in the network stays around a set point value equal to C.

7. STOP, but allow remodeling of the network (cells and synapses) such that if Vmem is above or below some network stabilizing Vmem value = [C +/- 'b'] (where 'b' is a variable bias on the membrane), then the system becomes activated, enabling plasticity (learning).

Inspection of the trabecular pattern of cancellous bone in the medulla of the adult femur, illustrates that extremely complex structures can develop from a simple bone model by utilizing scale-free, negative-feedback bio-rules guiding the formation of osteoblasts and osteoclasts. Osteoblasts add, and osteoclasts remove bone locally in response to mechanical strain that produces piezoelectric signals in the developing bone, and its collagen matrix. These signals affect the membrane potentials of the osteoblasts and osteoclasts. The whole system exhibits negative feedback and homeostasis in that the system shuts off when the bone maximally resists mechanical strain. This process produces a structure having maximum resistance to force using a minimum amount of bone [5]. Importantly, there is no global, top-down plan for development which could require an enormous amount of information. Instead, as in the construction of a complex termite mound, each bone-forming or bone-removing cell merely responds to local stress-strain-induced piezoelectric signals. The system remains active in the adult so that bone is able to become more or less sturdy as more or less force is applied to the bone. Exercise strengthens muscles, tendons, and bone. Osteoporosis results if the system fails to respond adequately to forces applied to the bone or bone loss outpaces bone formation.

Similarly, in synaptogenesis, signals from gap junctions, developing synapses, and local field potentials guide neurons so that network-balancing excitatory or inhibitory synapses are formed in a scale-free manner. Both bone and neocortical systems result in very complex structures via control of local cellular activity by electrical signals that, presumably, exert control over Vmem. If bio-rules for network formation operate at all scales, then they can result in networks that are fractals both spatially and temporally. This helps define a dynamical model in which a fractal small-world network can evolve in a self-organized critical (SOC) manner [20, 22]. The system undergoes self-organization through growth and decay of nodes, and synaptic modification.

The biological pseudo-code for neocortical development discussed above specifies an <u>iterated loop where the result of the output of the rule forms the basis for the next input to the rule</u>, in the same way a cellular automaton rule operates. Moreover, an iterated loop also suggests the Mandelbrot function, where K is some constant.

$$\longrightarrow F(z_{n+1}) = z_n^2 + K \longrightarrow$$

The Mandelbrot function forms an interesting fractal pattern if <u>z is a complex number [a +/- bi]</u>, where 'a' and 'b' are constants, and 'i' = $(-1)^{1/2}$, where these values fall within a given range.

In the real world of neuron circuits there are time-related action potentials which have <u>both magnitude and phase relationships</u>. These are easier to depict and evaluate if one uses complex numbers. This is the idea behind the possible requirement for complex-valued (C-V) neurons and neural networks. C-V neural networks have high functionality, plasticity, and flexibility[6].

Moreover, a <u>single</u> C-V neuron can learn input/output mappings that are <u>non-linearly separable</u> in the real domain without having to use higher dimensional spaces. A <u>C-V neuron can solve the XOR (eXclusive OR) problem</u>, a non-linearly separable problem[7]. Will a model of brain function that includes C-V neurons and networks minimize the number of neurons, energy requirements, and the volume of the neural network needed for calculations of XOR and similar functions? Additionally, does this also relate in any way to a possible requirement that neural networks are fractals in the sense of the Mandelbrot function[8]?

The exact biochemical mechanisms underlying this process are very complicated, and still being actively investigated. However, looking at cellular activity from the simpler perspective of Sundelacruz' work (with Vmem as a control signal), one can propose an overall <u>heuristic</u> based on 'rewiring' the fetal connectome so that its <u>information flow</u> changes from a random into a complex fractal pattern analogous to the <u>conversion of the intrinsic random information flow of A-graph #30 into the complex information flow of A-graph #110</u>.

Significantly, synaptogenesis might start with an <u>initial asymmetry</u> in synaptic structure where this initial asymmetry spreads at all scales throughout the developing neural connectome. We refer again to a rule proposed in a previous paper[1]:

<u>Asymmetry</u> + Computation (Randomization/Entropy) → Complexity.

Figure 4 below is a diagrammatic illustration of a randomly-connected network. Points or nodes along the circumference of the circle (not shown) are randomly interconnected with their immediate and very close neighbors, and these nodes are also connected at intermediate and longer distances, resulting in a random pattern (A-graph #30). Specifically, there are excess inter-nodal connections at intermediate distances (bold lines). The network is overconnected.

Figure 4—random network

The random network is not capable of complex computations because there is no constraint on information flow. To render the network complex, there have to be connection adjustments at short, intermediate, and long distances (at all scales) that parallel the conversion of A-graph #30 to A-graph #110.

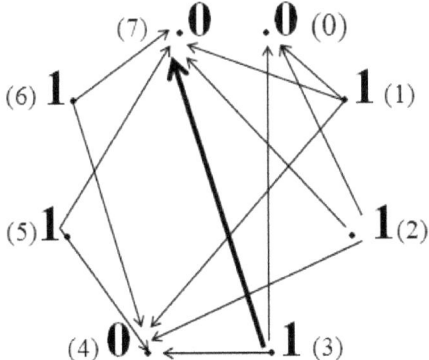

Figure 5

Eight nodes representing the distances at all scales in A-graph #110 are given the values of the digits of the binary number 110 (01101110). The binary digits are placed clockwise around the octagon. The numbers in parentheses represent distances along the abscissa in A-graph #110. Using the XOR function[1] we can form directed connections from binary digits with value '1' to binary digits with value '0.' The result is a connectome of eight nodes at one scale that has an information flow pattern like that of A-graph #110. Note, in particular, that there are four possible connections at an internodal distance = 4, but only 1 connection (in bold) is seen, giving a ratio of 1/4 or 0.25 (the ratio of actual/possible connections at that distance), thereby forming the 'valley' (V) at abscissa distance = 4 in A-graph #110. The ordinate of an A-graph is the ratio of actual to possible internodal connections at any given distance (abscissa).

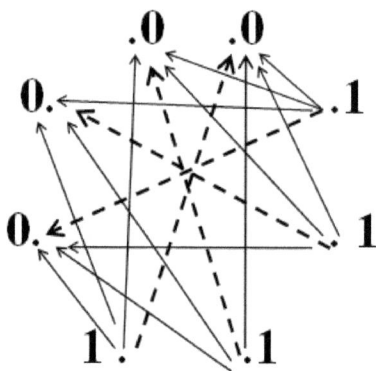

Figure 6

By contrast, Figure 6 illustrates the connectome at one scale described by A-graph #30. The binary digits around the graph are now those of the binary number 30 (00011110). Again, the binary digits of number 30 are placed clockwise around the octagon. The network is overconnected, especially at an intermediate internodal distance = 4 where the ratio of possible to actual connections (four dashed lines) is 4/4 =1 (actual/possible), representing the peak of the unimodal hill of A-graph #30. Alteration of these internodal ratios allows transformation of one A-graph into another. This is analogous to alterations in synaptic connectivity.

Various models of networks can be formed. For example, a crumpled piece of paper laid flat, shows a complex fractal pattern of intersecting ridges and valleys forming nodes[25]. If the ridges are marked as solid lines, and the valleys as dashed lines, then one can observe the formation of hubs of varying degrees, where a given hub can have different ratios of solid and dashed lines connecting to that node. If we let solid lines represent excitatory connections, and dashed lines represent inhibitory connections, then we can form a model of a random fractal network (connectome). Referring, then, to figures 5 and 6 above, we can alter the random connectome by adjusting interconnections at all scales to reflect the interconnection ratios seen in A-graph #110. Figure 7 below shows the model before any alterations.

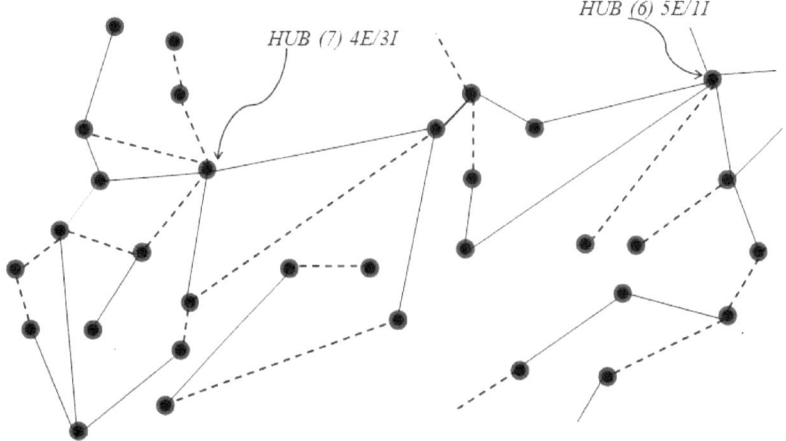

Figure 7

One region of a large crumpled piece of paper laid flat illustrates an overconnected random network. Ridges are marked with solid lines, and grooves with dashed lines Hubs of varying degrees are formed at the intersections of the solid and dashed lines. The region can be seen to consist of similar ridge-groove patterns at smaller and smaller scales (the thinner and larger the sheet of paper, the smaller the scales that can be formed). If we, then, add and remove excitatory and inhibitory connections such that the overall Vmem of the connectome reaches a dynamic equilibrium about the set point C as shown in Figure 2 above, then the information flow pattern is similar to the #110 A-graph. Two hubs are shown, one with seven connections (4 Excitatory and 3 Inhibitory), and another with six connections (5 Excitatory and 1 Inhibitory). Another instructive model can be formed using **Asymmetric Partitioning (AP)** of a region as shown in Figure 8 below.

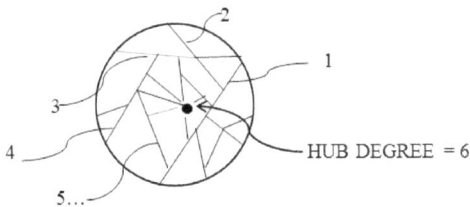

Figure 8

Draw line '1,' dividing the circle <u>asymmetrically</u> (i.e., not a diameter of the circle). Continue to partition asymmetrically with lines '2' and '3' and so forth so that the partitioning lines can sometimes pass, at random, through previously-formed intersections. Hubs of varying degrees are formed at all scales. Apply the <u>biologic pseudo-code</u>, noted above, so that the system approaches a homeostatic set point about 'C.'

Another example is seen in Figure 9 where we see the result of <u>sequential crack formation</u> in drying mud. Because the mud has asymmetries in thickness, composition, and surface curvature, each new crack <u>divides a remaining area asymmetrically</u> (once again allowing for random intersections at previously-formed intersections), such that after 'N' cracks the surface forms a fractal pattern at smaller and smaller scales. In contrast, if the original mud disk were perfectly symmetric with zero curvature, without asymmetries in the properties of the mud, and under no external forces, then the pattern of cracks might be simple such as a checkerboard.

Similarly, patterns of cracks (crazing) can be seen in pottery glaze, due to <u>initial asymmetries</u> in the thickness of the pre-fired glaze or variations in the <u>curvature</u> of the surface of the pot. The number of cracks increases with increased curvature of the surface of the pot. Once again, these cracks form a fractal pattern with nodes at the intersections of cracks.

Thus, in either the mud or pottery glaze models, <u>asymmetry plus randomization</u> results in complexity,[1,5,6]. In contrast, <u>symmetry plus randomization does not produce complexity</u>; but, rather, a regular, simple pattern.

In these models, adjustments in the number of connections at all scales, following the connection ratios in Wolfram A-graph #110, results in a small-world network as illustrated in Figure 9.

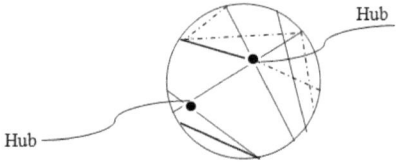

Figure 9—a small-world network
Figure 9 above illustrates a small-world network. The bold lines represent retained excitatory connections at intermediate distances. The excessive interconnections shown as bold lines in <u>Figure 4</u> above, representing connections at <u>intermediate</u> distances have either been <u>pared</u> or replaced by inhibitory connections or synapses (thinner dash-dot lines). The network in Figure 9

is not overconnected, and is capable of complex behavior. Note the paucity of connections and information flow at intermediate distances, similar to the A-graph of Wolfram cellular automaton #110 ('V' in Figure 1) Nodes along the circumference of the circle are not shown, but have connections to their nearest neighbors. Initially, a developing neural network is analogous to a random network as shown in Figure 4. As synaptogenesis proceeds, excitatory synapses at intermediate distances are eliminated and/or replaced by inhibitory synapses ('P' moves to 'V'), and short and long range connections are increased ('U' and 'V' move upward), resulting in a change from a random (A-graph #30) to a small-world network capable of complex computations (A-graph #110). See Figure 1.

Thus, flattened sheets of crumpled paper, and cracks in drying mud or pottery glaze can generate random network patterns. Nodes can be defined as the intersections of ridges and valleys in the former, and intersections of cracks in the latter. Asymmetric partitioning can also produce similar patterns. In Figure 10 below we see a pattern of cracks in mud or pottery glaze where an initial asymmetry N=1, 2, 3… is carried or projected into all scales producing a complex pattern of cracks. Importantly, as N increases, the **resulting fractal pattern is similar no matter where the first crack occurs.**

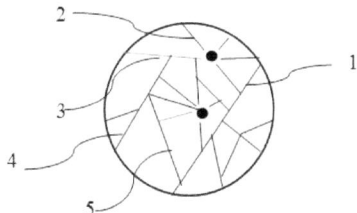

Figure 10

In Figure 10 the partitioned regions continue on at smaller and smaller scales, where each subsequent line in the diagram divides a remaining area asymmetrically. Place nodes at the intersections of lines, and observe that the intersections (black dots) can form hubs of various degrees.

Therefore, given any initial geometric or temporal asymmetry in neocortical synaptogenesis, if we let the process continue on at smaller and smaller scales, using the homeostatic biologic pseudo code noted above, with:

- paring of low Vmem neurons;
- the addition of inhibitory synapses that raise the time-averaged net Vmem of these neurons;

- the addition of excitatory synapses that lower the time-averaged Vmem of hyperpolarized neurons, while
- maintaining asymmetry throughout the process,

then this homeostatic process results in an ultra-small-world (fractal) network[7] capable of complex calculations seen in Wolfram cellular automaton #110.

It is important to note that in the crumpled paper, cracked mud or pottery glaze models, no matter where the first fold, crack or asymmetry occurs, one eventually ends up with a similar fractal pattern as the number of folds or cracks increases.

This means that the developing neocortex is probably not constrained to develop in only one way, i.e., along a single or unique pathway. Therefore, DNA does not have to code for a specific pathway that the developing neocortex must follow, because any initial asymmetry in synaptogenesis can lead to a similar (isomorphic) fractal small-world network. Thus, a significant decrease in the amount of DNA information is required for synaptogenesis.

All of this results in a scale-free small-world network with complex information flow that is isomorphic with information flow in the A-graph of #110 [10, 11, 12, 13, 14].

Figures 11 and 12 below illustrate how spindle-burst-induced electrical standing wave patterns in developing neurons are acted on by a putative biologic homeostatic mechanism, which guides the self-organization of a random neocortical network into a complex scale-free network analogous to the transformation of A-graph #30 into A-graph #110.

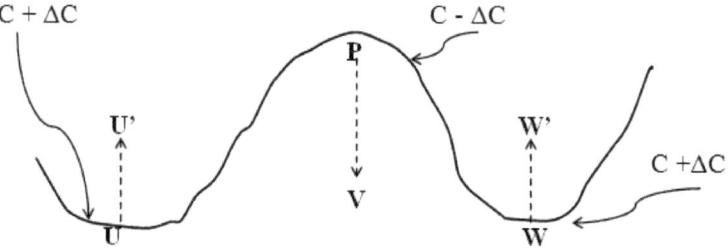

Figure 11—A-graph #30
Cross-section at one scale, and at one instant of time through the developing neocortex illustrating an instantaneous standing wave pattern of neuronal Vmem in response to random

spindle bursts. The central peak 'P' occurs where spindle burst waves are reinforced, and the valleys 'U' and 'W' where spindle burst waves interfere. The waves are fractals. As labeled, the central peak 'P' represents a region where the developing neurons are hypopolarized (C-ΔC = increased information flow), and the valleys 'U' and 'W' where they are hyperpolarized (C+ΔC = decreased information flow)[26]. Applying the homeostatic biologic pseudo-code to this system, results in neuron paring and excitatory and inhibitory synapse formation which 'smooth out' the peak and valleys of Figure 11, leading to Figure 12 below.

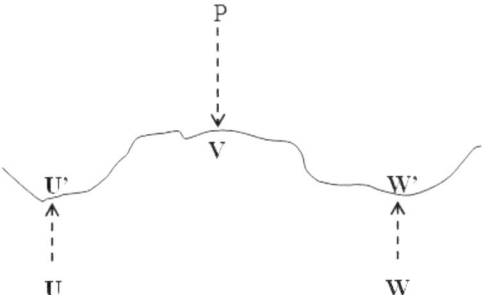

Figure 12— homeostatic 'smoothing out' of curve in Figure 11 yields A-graph #110
This figure illustrates that by neuron paring, and alteration of the ratios of excitatory and inhibitory synapses, the curve shown in Figure 11 has become smoother. Points have moved closer to the set point, C. Points U and W have moved up, and P has moved down to V (Figure 1). Vmem values at U' and W' have become more hypopolarized (increased information flow), and Vmem values at point V have become hyperpolarized (decreased information flow). Vmem regulates cortical connectivity[26]. If we equate hypopolarization with increased information transmission, and hyperpolarization with inhibition of information transmission, then we can relate the transformation of the curve in Figure 11 to the curve in Figure 12 as being equivalent to the transformation[21] of A-graph #30 to A-graph #110 (Figure 1).

Thus, conversion from random information flow in Wolfram asymmetry graph #30 to complex information flow in Wolfram asymmetry graph #110 is used as a heuristic to define a homeostatic model for synaptogenesis based on the effect of random spindle bursts on neuron transmembrane potentials. Several lines of evidence support the idea that Vmem controls neuron connectivity. This model helps explain the dynamic self-organization of the neocortex from a random network into a fractal small-world network capable of complex computation.

References

1. Goldberg, M. *Classification of Cellular Automata Using Asymmetry Graphs*. GRIN v341659, 2016.

2. Goldberg, M. Three papers: (1) '*Complexity Arising from Entropy Acting on Asymmetric Substrates.*' *Telicom* XII.22 June/July 1998, pp 38-40. Journal of the International Society for Philosophical Enquiry, ISSN: 1087-6456. (2) '*Complexity.*' *Telicom* XV.19, September 2002, pp 44-48. Journal of the International Society for Philosophical Enquiry, ISSN: 1087-6456. (3)'*Complexity and Randomness.*' *Telicom* XVI.3- June/July 2003 pp 65-67. Journal of the International Society for Philosophical Enquiry, ISSN: 1087-6456.

3. Wolfram, S., *A New Kind of Science* pp 488-489, ©2002 Wolfram Media Inc. Wolfram, Stephen, '*A New Kind of Science.*' Wolfram Media, Inc., May 14, 2002. ISBN: 1-57955-008-8.

4. Sundelacruz S, Levin M, Kaplan, DL. *Role of membrane potential in the regulation of cell proliferation and differentiation.* Stem Cell Rev. 2009 Sep; 5(3):231-46. doi: 10.1007/s12015-009-9080-2. Epub, 2009 Jun 27.

5. Frost, H. *The Laws of Bone Structure* (1964). Henry Ford Hospital Surgical Monographs, Charles C. Thomas publisher.

6. Hirose, Akira, Editor. *Complex-Valued Neural Networks.* Book: ISBN 978-1-118-34460-6 APRIL, 2013 Wiley-IEEE Press.

7. Aizenberg, I., et al, *Why we need complex-valued Neural Networks*, 2011, Manhatten College DOI: 10.1007/978-3-642-20353-4_1.

8. Szilagi, M., and Salik, B. *Neural Networks with Complex Activations and Connection Weights.* 1964 Complex System 8 (1994) 114-126 Departments of electrical and computer engineering, University of Arizona, Tucson, and Department of Electrical Engineering, California Institute of Technology Pasadena, CA.

9. *Scale-Free Network. Complex* network. Wikipedia.

10. Two Papers. (1) Stam, C., Jaap Reijneveld C. *Graph theoretical Analysis of Complex Networks in the Brain.* Nonlinear Biomedical Physics 2007 1:3 DOI: 10.1186/1753-4631-1-3. (2) Bullmore, E., Sporns, O. *Complex brain networks: graph theoretical analysis of structural and functional systems. Nature Reviews Neuroscience.* **10**, 186-198 (March 2009). doi:10.1038/nrn2575.

11.Theoden, I, et al. *Epilepsy in Small-World Networks.* Journal of Neuroscience 15 September 2004, 24(37) 8075-8083; DOI: https://doi.org/10.1523/JNEUROSCI.1509-04.2004.

12.Sima, M., Zochowski, M. *Measuring Predictability of Autonomous Network Transitions into Bursting Dynamics.* BMC Neuroscience2014 15(Suppl 1):P2.DOI: 10.1186/1471-2202-15-S1-P2.

13.Basalyga, G., Gleiser, P., Wennekers, T. *Emergence of Small-World Structure in Networks of Spiking Neurons through STDP Plasticity.* Volume 718 of the series Advances in Experimental Medicine and Biology pp 33-39.

14.Yingxi Lin et al. *Activity-dependent Regulation of Inhibitory Synapse Development by Npas4.* Nature 455, 1198-1204 (30 October 2008) | doi:10.1038/nature07319; September 2008.

15. Hanganu, I., et al. *Retinal Waves Trigger Spindle Bursts in the Neonatal Rat Visual Cortex.* Journal of Neuroscience 21 June 2006, 26 (25) 6728-6736; https://doi.org/10.1523/JNEUROSCI.0752-06.2006.

16. *Neural Development.* Wikipedia.

17. Bassett, D. et al. *Adaptive Reconfiguration of Fractal Small World Human Functional Networks.* PNAS: Vol. 103 no. 51 19518-19523 doi: 10.1073/pnas.0606005103.

18. *Seiche.* Wikipedia.

19. Gilson, M. et al. *Emergence of network structure due to spike-timing-dependent plasticity in recurrent neuronal networks.* Biol. Cybern. (2009) 101:427–444 © Springer-Verlag 2009. doi 10.1007/s00422-009-0346-1.

20. Watanabe, A. et al. *Fractal and Small-World Networks Formed by Self-Organized Critical Dynamics.* J. Phys. Soc. Japan. 84, 114003 (2015) [10 Pages] http://dx.doi.org/10.7566/JPSJ.84.114003.

21. Butz, M, et al. *Homeostatic structural plasticity increases the efficiency of small-world networks.* Front. Synaptic Neurosci., 01 April 2014. https://doi.org/10.3389/fnsyn.2014.00007.

22. Le Magueresse, C., Monyer, H. *GABAergic Interneurons Shape the Functional Maturation of the Cortex.* http://dx.doi.org/10.1016/j.neuron.2013.01.011.
23. Cohen, R., Havlin, S. *Scale-Free Networks Are Ultrasmall.* Physical Review Letters Vol. 90 Number 5 (week ending Feb. 2003). Minerva Center And Department Of Physics, Bar-Ilan University, Ramat-Gan, Israel.

24. Complex Networks. Wikipedia.

25. Andresen, C. A. *Properties of Fracture Networks and Other Network Systems.* 2008 Thesis for PhD, Norwegian University of Science and Technology, Dept. of Physics.

26. Özcucur, N. et al. *Membrane Potential Depolarization Causes Alterations in Neuron Arrangement and Connectivity in Cocultures*. Brain and Behavior. Nov. 2014. Doi: 10.1002/brb 3.295

27. Bullmore, E., Sporns, O. *The Economy of brain network organization.* Nature Reviews Neuroscience **13**, 336-349 (May 2012) | doi:10.1038/nrn3214.

****